Mixtures and Solutions:
The Sugar in the Tea

by Emily Sohn and Joseph Brennan

Chief Content Consultant
Edward Rock
Associate Executive Director, National Science Teachers Association

Chicago, Illinois

Norwood House Press
PO Box 316598
Chicago, IL 60631

For information regarding Norwood House Press, please visit our website at
www.norwoodhousepress.com or call 866-565-2900.

Special thanks to: Amanda Jones, Amy Karasick, Alanna Mertens, Terrence Young, Jr.

Editors: Jessica McCulloch, Michelle Parsons, Diane Hinckley
Designer: Daniel M. Greene
Production Management: Victory Productions, Inc.

Paperback ISBN: 978-1-60357-312-2

The Library of Congress has cataloged the original hardcover edition with the following call number: 2011011443

282R—082015
Printed in ShenZhen, Guangdong, China.

CONTENTS

Note to Caregivers:

Throughout this book, many questions are posed to the reader. Some are open-ended and ask what the reader thinks. Discuss these questions with your child and guide him or her in thinking through the possible answers and outcomes. There are also questions posed which have a specific answer. Encourage your child to read through the text to determine the correct answer. Most importantly, encourage answers grounded in reality while also allowing imaginations to soar. Information to help support you as you share the book with your child is provided in the back in the **Additional Notes** section.

Words that are **bolded** are defined in the glossary in the back of the book.

Mix It Up!

You may not realize it, but you are already a **chemist.** Every time you mix things up or watch them change, that's science in action. Maybe you've put fruit and milk in your breakfast cereal. Or you might have watched burning wood turn to ashes. Bean soup and banana bread are works of science, too. In this book, you will learn how materials mix—or don't. As you combine objects, you will see how they change. You will also see how they stay the same. You will use what you learn to solve a sweet mystery.

How Much Sugar Can You Put in Tea?

Your older brother has poured two drinks. There is a cool glass of iced tea for you. He is having a glass of steaming hot tea. Both drinks are unsweetened. So, he gives you the sugar bowl. "Please do not put too much sugar in our drinks," he says. "One teaspoon in each should be enough." Then he leaves the room.

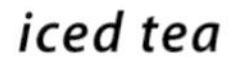

iced tea

hot tea

When your brother comes back, he looks angry. "You put more sugar in your iced tea than you put in my hot tea!" he says.

You protest. "Our cups are about the same size," you say. "And I used the same amount of sugar in both. Honest!"

"Then why is there sugar at the bottom of your drink," he asks, pointing at the visible grains, "but there is nothing at the bottom of mine?"

You are sure that you put the same amount of sugar in both drinks. How can you prove you are telling the truth?

The next two pages offer four things you might try to do to convince your brother. All four are types of demonstrations. As you think about which idea is best, consider these questions:

- Will the result of each demonstration prove you are telling the truth?
- What **properties** of the drinks are you testing?
- What else do you need to learn in order to decide what to do?
- Will you and your brother still be able to drink the tea after you're done experimenting?

Idea 1: Let Them Settle

Let the glasses sit on the counter for an hour. Don't stir or touch them. Then see if you can see any sugar to measure.

Idea 2: Start Over

Pour two new mugs of tea, one with sugarless iced tea and one with hot tea. Add a spoonful of sugar to the hot tea and stir. If all of the sugar disappears, add another spoonful. Do this until some sugar settles at the bottom of the mug. Count how many spoonfuls you used. Repeat with the iced tea.

Idea 3: Filter Them

Get two pieces of filter paper, such as coffee filters. Over a sink, pour the iced tea through one filter. Pour the hot tea through the other. Are you able to trap sugar to measure?

Idea 4: Turn Up the Heat

Set both drinks in a sunny spot. Over time, all of the liquid will turn into a gas and escape into the air. Depending on how hot and humid it is, this could take weeks. Does any sugar remain in either glass?

DISCOVER ACTIVITY

Materials
- yarn
- 2 plastic cups
- lid from a large jar
- salt
- spoon

Make a Salt Column

You want to get the sugar out of your drinks. But first, try this investigation with salt. It will help you understand how solids and liquids mix.

Fill two cups with hot tap water. Put the cups in a place where they can sit out of the way for a week or two. Use the spoon to add a little salt to both cups. Stir. Keep adding salt and stirring until you see some salt grains at the bottom of the cups.

Now twist a few strands of yarn together to make a thick string. Place one end of the string in each cup. The string should connect the cups.

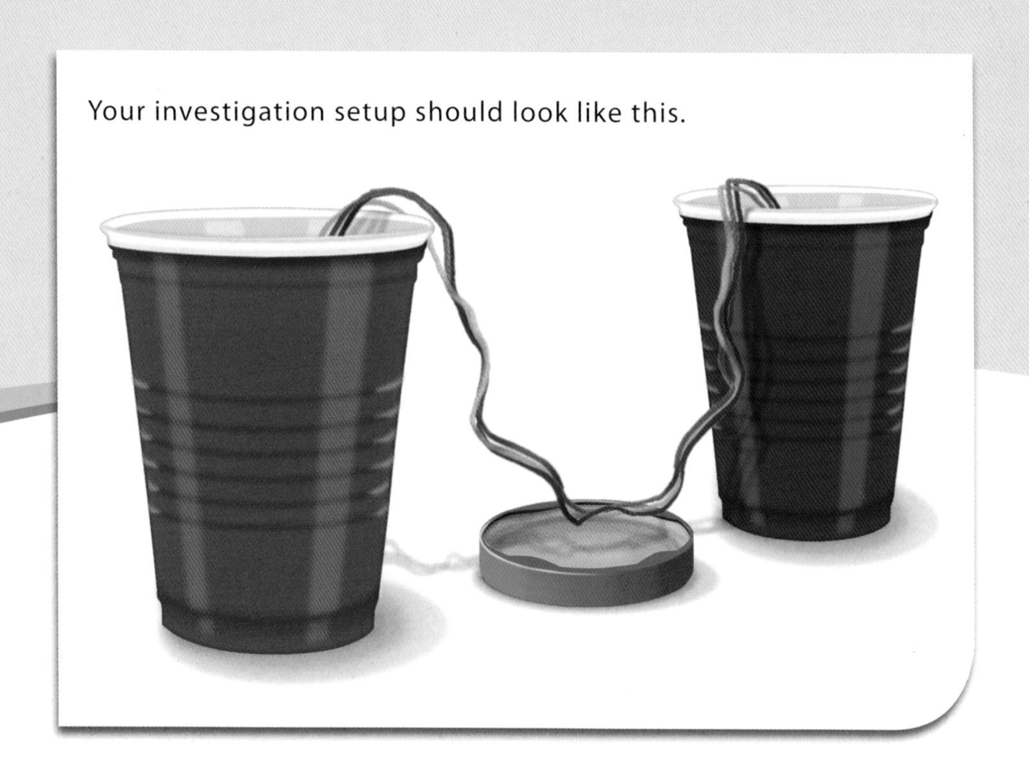
Your investigation setup should look like this.

Allow the yarn to sag in the middle between the cups. Place the jar lid under the sagging yarn. The lid should face up so that it makes a little cup.

After a few minutes, water from the cups will travel along the yarn. Then, it will begin to drip into the jar lid. Watch the cups and the yarn every day for 10 to 14 days. Record what you see.

What did you observe? Can you explain your observations? Do not worry if you cannot fully explain what's happening. As you read this book, you will learn the science you need to know. That will help you make sense of this activity.

What Are Objects Around Us Made Of?

The iScience Puzzle involves sugar, tea (which is mostly water), and one more addition. There's also the air you're breathing, right? All of these things have some similarities. For example, you can hold or touch each one with the palm of your hand. To touch air, you simply have to wave your hand in the air. (You could also blow up a balloon and hold it to help you visualize the air. But the rubber of the balloon will be between you and the air.)

You can see lemons in water. Both lemons and water are visible forms of matter.

They have many differences, too. Water pours. Air is invisible. Sugar disappears when you mix it into certain drinks.

Each of these is a form of **matter.** And each form of matter behaves in unique ways.

A World of Stuff

List everything around you right now. You might see objects that are big or little, hard or soft, round or square. As different as they are, all **substances** on Earth are types of matter. And most matter is made of **atoms.**

Some substances are made of only one kind of atom. These are called **elements.** Other substances are made of more than one kind of atom. Combinations of atoms are called **molecules**. A molecule can be made of atoms from only one element. A molecule of oxygen, for example, is made of two oxygen atoms. A molecule can also be made of atoms from more than one element. The water molecule and the sucrose molecule in the pictures below are this kind of molecule.

A molecule of water is made of 2 hydrogen atoms and 1 oxygen atom. Its **chemical formula** is H_2O. A molecule of table sugar, or sucrose, is made of 12 carbon atoms, 22 hydrogen atoms, and 11 oxygen atoms. Its chemical formula is $C_{12}H_{22}O_{11}$.

Water molecule:
red = oxygen atom
gray = hydrogen atom

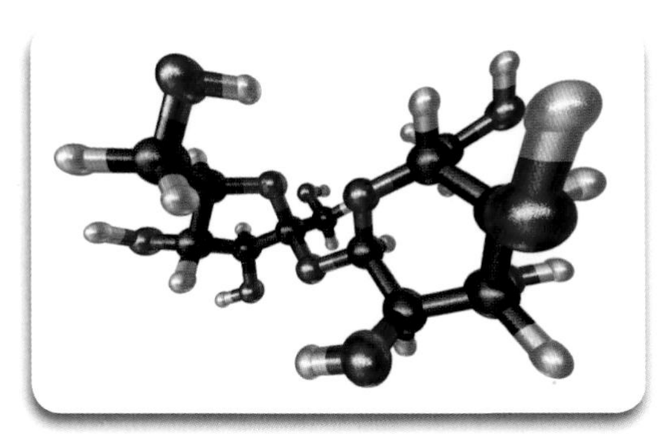

Sucrose molecule:
red = oxygen atom
gray = hydrogen atom
black = carbon atom

Each kind of atom and molecule behaves in a unique way. Their properties affect what happens when you put them together. Sometimes, nothing happens. Sometimes, a lot happens.

Water and sugar blend so well that the sugar becomes invisible. Can you name two ways in which water and sugar are different from each other?

Doing the Mix

When you put sugar in your drink, you made a type of **mixture.** The solid grains of sugar spread out into the liquid. You couldn't see all the sugar. Still, both the sugar and the tea stayed the same. Their properties did not change.

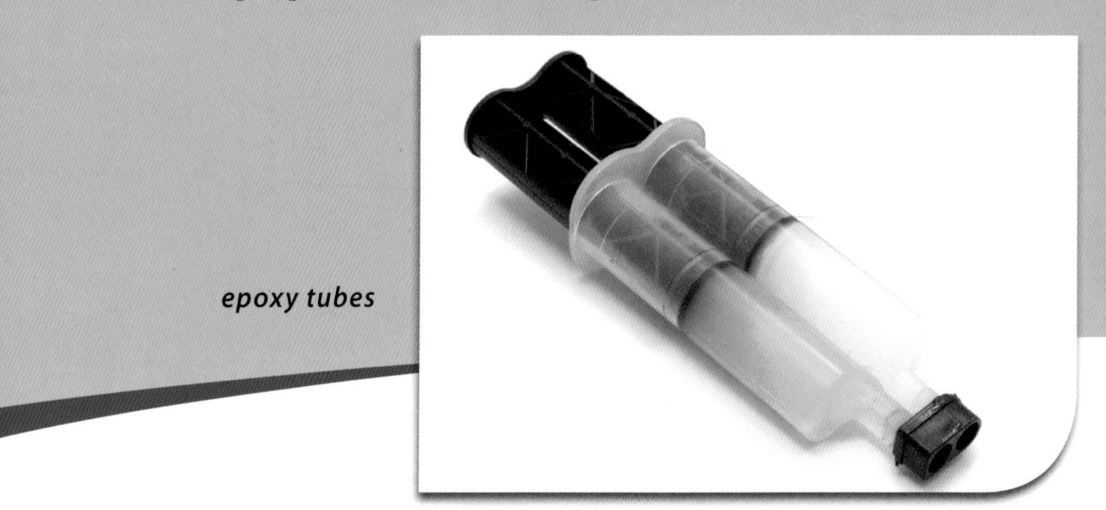

epoxy tubes

Sometimes, substances change when they combine. A type of glue called epoxy is one example. It comes in two separate tubes. Each tube contains a different liquid. After mixing, the substances join to form a completely new **compound.** They become one sticky liquid. Together, the liquids quickly turn into a strong solid that holds things together. Once the solid has formed, you cannot separate it back into its parts.

Have you ever made a cake? You start with many ingredients. You stir them all together to make a mixture called cake batter. Then you pour the batter into a cake pan and put the pan in a hot oven. When the cake is fully baked, it is a darker color than the batter was. You cannot pour the cake like you poured the batter. You cannot separate the cake back into its ingredients. A new mixture of substances, a cake, has formed.

Think about things you've mixed while cooking or doing an art project. Are they more like epoxy or more like sweet tea?

What would happen if you let this glass of muddy water sit untouched for a day?

Making Sense of Mixtures

Gather a pile of little things. You might find coins and keys. Add toy cars and paper clips. When things like this get jumbled, it is easy to pick them apart again. That means they are a mixture.

Now, consider something messier, such as mud. Mud is a combination of dirt and water. It takes a little work to find out if mud is a mixture or a compound. You could stir up some muddy water in a glass. Then you could check on it over the course of a day. The dirt would likely settle to the bottom. The water would stay on top. This shows that mud is also a mixture.

Do you think iced tea is a mixture or a compound? What about hot tea? Could you use the same method you used with mud in a glass to find out?

The front-end loader is picking up loads of dry sulfur. Heating iron with sulfur forms a compound called iron sulfide.

The Art of Mixing

Some materials can mix in more ways than one. Iron and sulfur are an example. These are common substances in factories. Combining tiny bits of iron with powdered sulfur makes a mixture. You could remove the iron bits by picking them out. To remove them even more quickly, you could use a magnet to attract and pull out the iron bits.

Something else would happen if you mixed iron pieces with sulfur and then heated the mixture. You would produce a compound called iron sulfide. It is also called pyrite, or fool's gold. It looks like a hard chunk of crystal.

pyrite, also known as fool's gold

Iron sulfide does not look like its parts. Once iron and sulfur form a compound (pyrite), you can no longer separate them. Even a magnet will not pick out the iron.

There is a single reason for all of these changes. A **chemical reaction** has occurred. The iron and sulfur have combined and changed, making a new substance. Notice that adding heat made all the difference.

Think about the sugar in your drinks. Do you think a chemical reaction could explain why the sugar seemed to disappear?

Pyrite is called fool's gold because it looks a lot like gold but is much less valuable. Some jewelry is made from pyrite. And the Incas of South America made mirrors from the shiny crystals.

Leave an oil-and-vinegar mixture alone long enough and the oil and vinegar will separate into layers. If you left a piece of pyrite alone long enough, do you think it would separate into iron and sulfur?

Pulling Apart the Pieces

In some cases, it is as easy to separate parts as it was to mix them. Look at a bottle of salad dressing that contains oil and vinegar. When the bottle sits still, oil and vinegar drift into two layers.

Other mixtures require work or tools to pull them apart. You learned how magnets pull iron out of mixtures. In other cases, filters can help. Some people make coffee by pouring hot water over tiny pieces of coffee called grounds. A filter allows water mixed with coffee to pass into a cup. But it keeps the larger coffee grounds out.

Could you use a magnet to pull sugar out of your drinks? Could you use a filter to do it?

The blade of this knife is made of stone.
The handle is made from a piece of antler.

CONNECTING TO HISTORY

The Bronze Age

Humans have been using tools for millions of years. During the Stone Age, people shaped tools out of found objects. They used sharp stones for cutting. They probably used sticks to dig for roots they could eat. This period started about 2.5 million years ago. It lasted until about 10,000 years ago.

Then, people started working with copper. This metal could be mined out of rocks. People could make new kinds of tools with copper. But copper tools were not very hard or sturdy. Stone tools often worked better.

A big shift came about 5,000 years ago. That's when people in the Ancient Near East and other places started to experiment with metals. They were able to improve copper by mixing it with other metals. They did the mixing under high heat. Combining copper with tin made a hard metal called bronze. This was the start of the Bronze Age.

Later, people mixed copper with the metal zinc. The result was brass. This metal was harder and stronger than copper. It had a lot of other useful properties. It was colorful, resistant to wearing away, and easy to shape.

How are iced tea, hot tea, and sugar similar to these metals? How are they different?

bronze door knocker

brass door knocker

What Are Elements?

It can seem hard to know what will happen when you mix things together. To plan studies and predict results, scientists turn to atoms.

In this factory, the gases that make up the air we breathe are separated from one another. Then they are put to separate uses.

Atoms make up all kinds of matter. And atoms come in many forms. Remember that a substance made of just one type of atom is called an element. Chemists can separate compounds into their elements. And they can combine elements in new ways. These ideas are at the root of medicine, food science, and more.

This lithium atom has 4 neutrons (blue) and 3 protons (red) in the nucleus. It also has 3 electrons (black). The electrons don't follow a set path. Rather, they travel in a random motion around the outer edges of the atom.

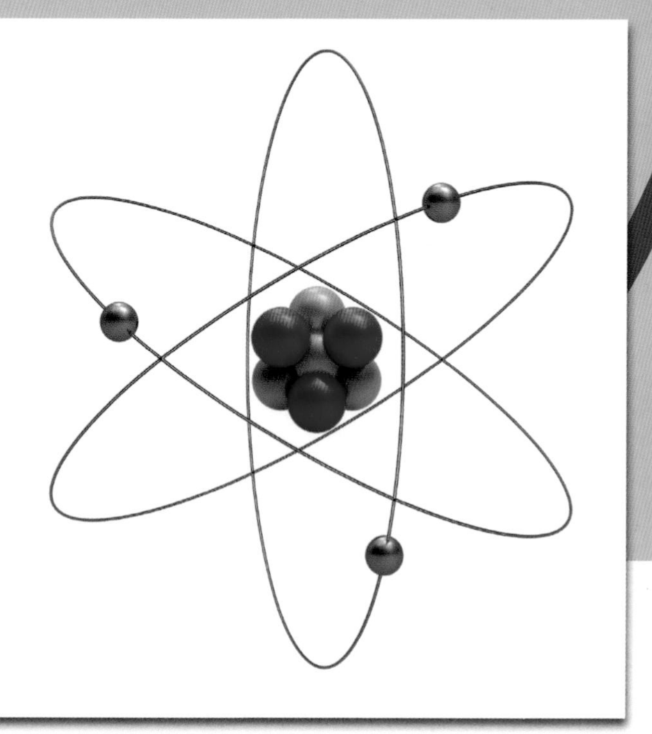

Smaller than Small

Atoms are really tiny. But atoms have parts, too. In the early 1900s, scientists realized that electricity comes from **electrons.** These particles have negative charges. They move around within an atom. Or they speed along a path to light up a bulb.

Each atom also has a positive charge in its center, or **nucleus.** The nucleus contains **protons.** Protons carry a positive charge. The nucleus also has **neutrons,** which have no charge. Overall, the atom has no charge because the number of protons equals the number of electrons.

Most of the mass of the atom is in the nucleus. Why do some elements have more mass than others? Each element has a different number of protons and neutrons. Oxygen atoms, for example, have more mass than lithium atoms because each oxygen atom has more mass in its nucleus.

The melting point of water is 32°F (0°C).

Elements act in unique ways. For example, gallium is a hard metal. But the heat of your hand will melt it. Chromium, on the other hand, stays solid until about 3,300°F (1,816°C). This property is called the melting point. It is the temperature at which a substance turns from a solid to a liquid.

Color is another property of elements. So are hardness and density. Some elements are good at conducting electricity and heat. Some freeze at extra-low temperatures.

Water is a compound made of hydrogen and oxygen. What are some of its properties?

Little Mixers

Atoms are not usually found alone. Instead, they tend to stick together. Atoms combine into molecules. You cannot see molecules, but you can see compounds that contain huge numbers of molecules.

Compounds contain at least two kinds of elements. Sugar is a compound. It is made of three elements: hydrogen, oxygen, and carbon.

Table salt is made from two dangerous elements. Sodium and chlorine are not safe for eating, but salt is.

Atoms can link up in many ways. Results can be helpful, messy, and even a little scary. Sodium atoms are one example. Sodium combines with chlorine to make the salt we eat. But by itself, chlorine gas can be harmful to someone who breathes it in. And sodium atoms react with water to make flames. This can cause an explosion.

What Is a Chemical Reaction?

Just as elements can chemically combine with other elements, compounds combine with other compounds. To act this out, link arms with a friend. Pretend that each of you is an atom linked with a different kind of atom. You represent a molecule in a compound.

Iron and water combine to form a new substance—rust. The new substance cannot be changed back into iron and water.

Now break connections with your friend and join with two other classmates. You have just shown what happens when atoms in molecules break apart. They rejoin with other atoms to form new molecules. These new molecules make up new compounds.

Atoms react with each other through their electrons. These tiny particles move around the outer parts of atoms. An atom can take electrons away from other atoms. It can share electrons with other atoms. Or it can give up some of its own electrons. In all cases, the atoms become joined as compounds.

Does the sugar in your drinks share any electrons with the liquid it's in?

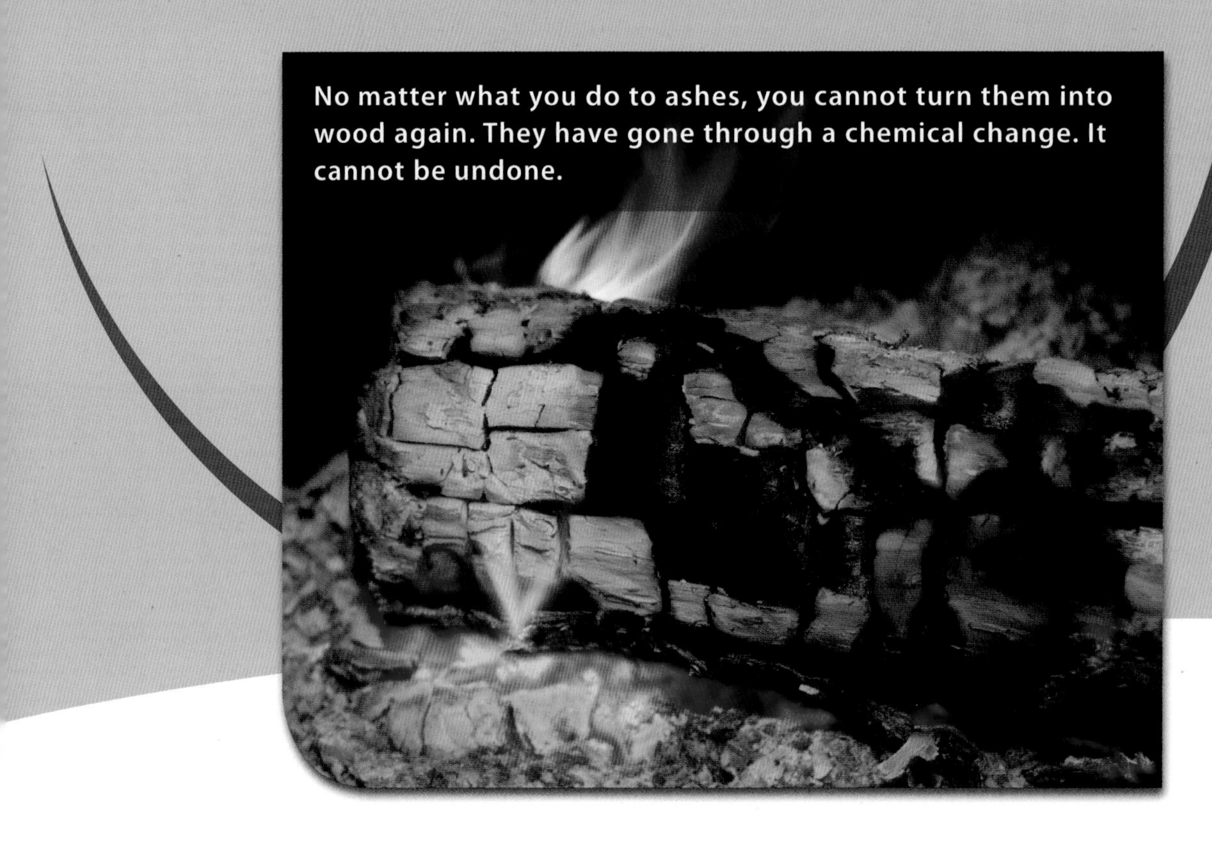
No matter what you do to ashes, you cannot turn them into wood again. They have gone through a chemical change. It cannot be undone.

A World of Change

Our days are filled with things that change. And change can come in many ways. On a hot day, you might put water in the freezer to make ice. This is a **physical change.** No new substance has formed. Even though it's frozen, the ice is still water. Ice can melt and turn back into water again. Only the shape or form has changed.

On a cold day, you might burn wood in the fireplace. Wood reacts with heat and oxygen in the air. Ashes form. This is a **chemical change.** In a chemical change, at least one new substance forms. It is not possible to change the ashes back to wood again.

Does mixing sugar into water cause a chemical change or a physical change? How do you know?

You are probably familiar with more chemical reactions than you realize. Adding heat to cookie dough causes a chemical reaction that hardens the dough into cookies!

Mysterious Signs

Think back to the iScience Puzzle. Has some sugar gone missing? To solve cases like these, you want to look for signs of how things have changed.

Most chemical changes are obvious. Some reactions give off light and heat. Some give off gases. There can be a smell or a sound when this happens. The new compound that forms does not look or act like its parts. It has different properties. Chemical changes either take in or give off energy. That can make things warmer or cooler.

Do you see signs of a chemical change in the iced tea or hot tea?

Nothing to Lose

Mass might help you figure out the puzzle. Mass is the amount of matter in an object. A French scientist named Antoine Lavoisier was working in 1778. He burned hydrogen and oxygen inside a sealed glass container. The chemical reaction left drops of water on the inside of the glass. (After all, water molecules are made of hydrogen and oxygen.) Lavoisier measured the masses of both gases before and after the reaction. There was less gas at the end. But the water that was created equaled the mass of the gas lost.

Did You Know?

Fireworks are full of chemical reactions. Their colors depend on the chemicals that are used to make them. When sodium salts burn, we see yellow. Calcium looks orange. Barium and copper appear green and blue. Magnesium and aluminum give off an intense white.

Lavoisier's experiment showed that no mass was gained or lost during the chemical reaction. This concept became the Law of Conservation of Mass. To conserve something means to save it. In a chemical reaction, all the atoms in the original substances are conserved, or still there, in the new substances formed in the reaction.

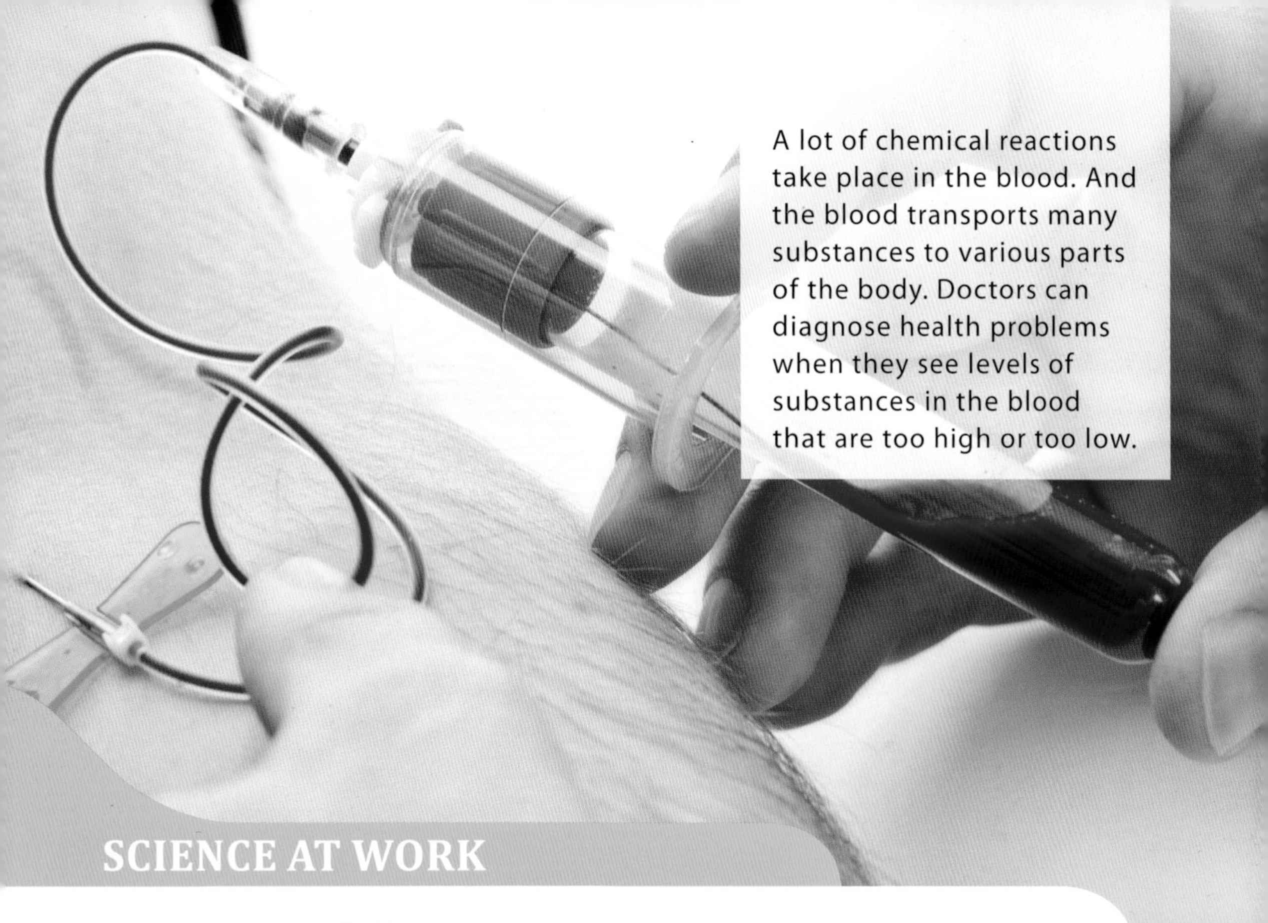

A lot of chemical reactions take place in the blood. And the blood transports many substances to various parts of the body. Doctors can diagnose health problems when they see levels of substances in the blood that are too high or too low.

SCIENCE AT WORK

Doctors and Nurses

Doctors and nurses have to study chemistry in school. That's because our bodies are full of compounds. They include vitamins, hormones, metals, and gases.

Chemical reactions happen in the human body all the time. We digest food. We breathe air. We use our muscles to run and lift things. With each action, millions of reactions occur. Doctors and nurses need to check our body chemistry. That way, they can make sure everything is working right. When reactions go wrong, we can get sick.

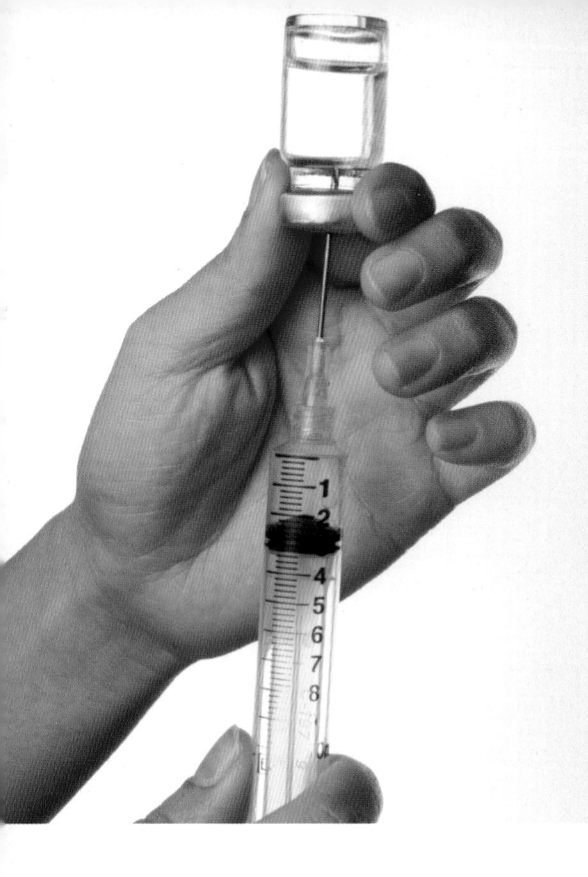

A nurse fills a needle with the right amount of medicine for a patient.

Medicines are chemicals, too. Doctors and nurses need to know how those chemicals work in the body. If they give a wrong medicine or the wrong dose (amount) of a medicine, they can harm or fail to help a patient.

It gets even more complicated! Every person's body chemistry is different. A person's body weight affects how a drug will work in that person. Some medicines affect how other medicines work. And medicines come in different strengths. Doctors and nurses use mathematics and formulas to figure out how much medicine is safe for each patient.

Children often get smaller amounts of medicine than adults get. Sometimes, they get different medicines altogether. Can you explain why?

This mixture of cement, gravel, sand, and water will form a hard concrete sidewalk when it dries.

Next time you're in a car or bus, look at the road below. Atoms help make your ride bumpy or smooth. That's because of the way mixtures combine—or don't. Loose gravel and sand never combine chemically. So the ride can be rough. Cement is another story. It is a blend of lime, clay, and other substances. These ingredients are finely ground into a powder. Mixed with gravel, sand, and water, they harden into concrete.

Concrete cannot be pulled apart into its pieces. Why?

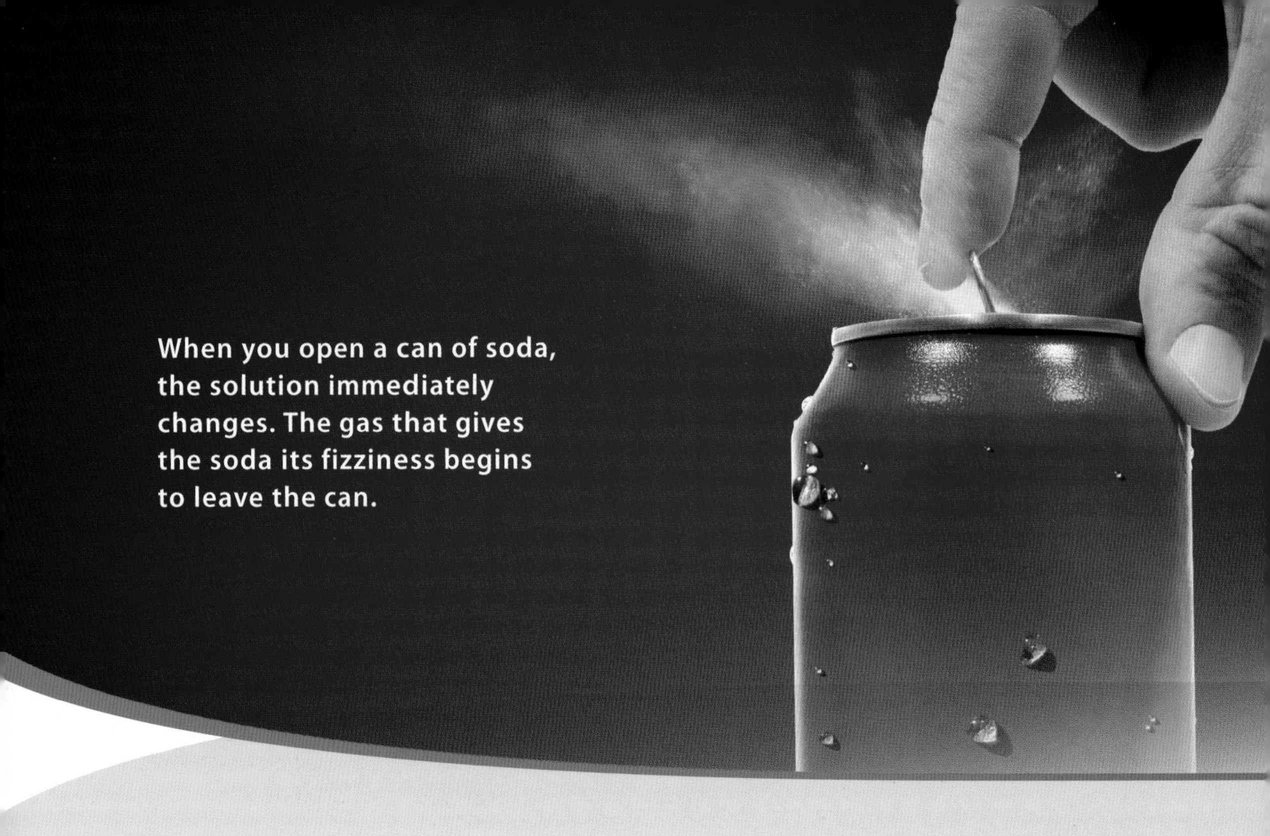

When you open a can of soda, the solution immediately changes. The gas that gives the soda its fizziness begins to leave the can.

Think back to the Discover Activity. You made salt seem to disappear into water. This kind of mixture is called a **solution.** The salt blended so well into the water that you could no longer see it.

Gasoline and soda drinks are also solutions. Making them involves physical changes. That means that they can be separated again. When you open a can of soda, you hear a fizz and bubbles form. Both clues show you that gas is leaving the solution.

Is sugarless iced tea a solution? Does adding sugar change your answer?

Brown sugar crystals are larger than white sugar crystals. Both kinds, however, get pulled apart in liquid.

Disappearing Act

In solutions, you can make things vanish. Think about the iScience Puzzle. If you could make atoms really big, you'd see that sugar is made of tiny crystals. The molecules are linked in precise patterns.

When you add sugar to tea, water molecules pull the crystals apart. The sugar mixes in the water in a separated form. The molecules then spread throughout the liquid. They have **dissolved.** Why do you think dissolving makes sugar seem to disappear?

Stirring sugar into tea can help the sugar dissolve faster.

Faster Reactors

You're really excited to drink your iced tea. You can make sugar dissolve more quickly. One way is to stir it. This makes each sugar crystal come into contact with more water molecules. Do you think stirring also allows more sugar to dissolve in a drink? How can you find out?

Heating things up might help, too. Think back to the iScience Puzzle. Iced tea is colder than hot tea. How could you test whether temperature affects how much of something dissolves?

Time to Concentrate

Your brother claims you put more sugar in your own drink. If the amount of liquids were the same, that would mean yours is more concentrated. Liquid with more of a substance dissolved in it has a higher **concentration** of that substance.

The longer you leave the bag in the cup, the more concentrated the tea will become.

Sometimes, solutions look different when they are highly concentrated. Watch what happens when you leave a tea bag in hot water. Over time, the tea gets darker. Its concentration rises.

Some people like one spoonful of sugar in their drinks. Some prefer two, or none. If the kind and amount of liquids were the same, which would have the higher concentration of sugar?

Think back to the iScience Puzzle. How do the concentrations of sugar in the two drinks compare?

Did You Know?

The Dead Sea is a lake that lies on the borders of Israel and Jordan. It is nearly nine times saltier than the ocean. In the Dead Sea, there is about one salt molecule for every three water molecules. People do not sink into this water. Instead, they float like a cork. The Dead Sea is so salty that it burns the eyes and skin of people who swim in it.

Which do you think takes longer to dissolve, sugar cubes or loose sugar?

The size of crystals affects how things dissolve, too. Coarse sugar has bigger crystals than fine sugar has. Sugar cubes are extra-big. Which type of sugar do you think you'd have to stir the longest? Might crystal size affect the amount of sugar that will dissolve into tea? How could you find out?

You don't need to swim to stay afloat in the Dead Sea. That leaves your arms free for other things, like holding a newspaper.

Do you think all of this sugar would dissolve into one cup of tea?

You can make tea really sweet. But there is a limit. At some point, the spaces between water molecules fill up. The sugar molecules run out of room and fall to the bottom of the cup. When no more solid can dissolve in a liquid, a solution is **saturated.**

Some compounds dissolve into solutions better than others. Only a small amount of lead chloride will dissolve into water. But water can hold a lot of zinc chloride. The two compounds have different saturation points.

How would you know when a drink becomes saturated with sugar?

How Do You Separate a Solution?

Mixing up a solution is as easy as squeezing chocolate syrup into milk. But separating the parts can be tougher. The task is hardest if a chemical change has happened. Then, it takes a new chemical reaction to reverse the process. If a physical change has happened, the task is still hard because the parts are mixed so well.

Could a paper coffee filter collect the sugar molecules from a cup of tea?

Think back to the iScience Puzzle. Could you use magnets or your fingers to pull the sugar out of the drinks?

Vaporize

To separate solutions, you might need to pull some other tricks out of your sleeve. **Evaporation** is one idea. It explains why water slowly disappears from a glass in a sunny spot. Energy from the Sun causes water molecules to move faster. They also move farther apart. Some of the molecules turn into a gas, called water vapor.

When you leave wet clothes on a clothesline, the liquid water in the clothes turns into water vapor and floats away. Dry clothes are left behind.

Now think back to the Discover Activity. What happened to the salt molecules that were dissolved in the water? What happened to the water? What can you conclude?

Weather depends on evaporation. First, water evaporates from oceans, lakes, rivers, and other places on Earth. In the sky, the water vapor molecules come back together and form clouds. When the water vapor gets too heavy in the clouds, it falls as rain or snow.

Some liquids evaporate at faster rates than others. Substances that are solid, such as sugar, don't evaporate.

Think back to the last idea in the iScience Puzzle. What would happen if you left the drinks in a sunny spot?

Dew forms when water vapor in the air condenses on cool blades of grass.

What Goes Up

Say you did set a glass of iced tea in a sunny place. Some of the water in it would evaporate. But the outside of the glass might also become wet. This happens when warm water molecules in the air hit the cold glass. The water molecules lose energy, slow down, and move closer together. As a result, they turn from a gas into a liquid. This is called **condensation.**

Clouds contain water that has condensed. Think back to the previous page. When water evaporates, it rises. Up high in the air, it cools. Cooler water molecules slow down. The vapor collects on tiny pieces of dust and forms clouds. The vapor condenses back into liquid water. When water in clouds gets too heavy, it falls as rain or snow.

Evaporation and condensation are opposites. But they can work together.

Drink It

Together, evaporation and condensation can do some useful things. Distillation uses both to get substances out of solutions. It can be used to make water safe to drink.

"Desalinate" means "to remove the salts from" something. In this desalination plant, salt is taken out of seawater to make the water safe to drink.

Much of the water on Earth is too salty to drink. To distill it, some factories heat salty seawater until it evaporates. That leaves the salt behind. Water vapor is collected in another container. Then, it gets cooled until it turns back into a liquid. Pure water is left behind.

How could you use distillation to solve the iScience Puzzle? Remember that you want to measure the sugar in each glass. But you still want to be able to drink your iced tea!

SOLVE THE iScience PUZZLE

You are still trying to prove to your brother that you added the same amount of sugar to both drinks. What have you learned that helps you choose from the four ideas?

Idea 1: Let Them Settle

Let the glasses sit on the counter for an hour without stirring. Measure any sugar you see.

Idea 2: Start Over

Pour two new mugs of tea, one hot and one cold. Add sugar to each mug and stir until no more sugar dissolves. Count the number of spoonfuls you added to each mug.

Idea 3: Filter Them

Pour the drinks through a filter into the sink. Will the paper trap the sugar? If so, then you can measure the sugar.

Idea 4: Turn Up the Heat

Put both drinks in a sunny spot. Wait for the liquid to evaporate. Is there any sugar left behind to measure?

In this book, you have learned that solutions are special types of mixtures. They do not separate easily. Molecules of sugar stick between molecules of water. You could wait and watch all day for them to settle. But idea 1 will not work. Dissolved sugar molecules are too small to get caught in a filter. So idea 3 is out, too.

Idea 4 could work. But it might take weeks for the liquid to evaporate. After all that, you wouldn't be able to drink your tea. The best idea is to perform a new experiment, like idea 2 suggests.

You could pour a fresh glass of iced tea and the same amount of hot tea. With your brother watching, you could prove that more sugar dissolves in water when the liquid is hot. After all your hard work, you could invite some friends to join you for a tea party!

When you're done investigating sugar solutions, throw a tea party!

The next time you go to a grocery store, walk down each aisle. Read labels on foods, cleaning supplies, medicines, and other products. Look at the ingredients. Then decide: Are they mixtures or solutions? Do you recognize any compounds on the labels?

With the help of an adult, try some kitchen experiments. Can you separate products into their parts? Start with granola bars. Move on to orange juice. Next, mix up some ingredients from the pantry. Are the changes that happen chemical or physical? Can you un-mix what you've mixed up?

Finish with baking soda and vinegar. Conduct this experiment outside or in a sink. Wear glasses or goggles for eye protection, and an apron. Add vinegar to a 16 fl. oz. (473 mL) bottle so that it is 1/3 full. Measure 3 heaping teaspoonfuls of baking soda into a cup and pour it into the vinegar. As you mix, stand back. This demonstration is very safe and will not explode. But the chemical reaction might surprise you!

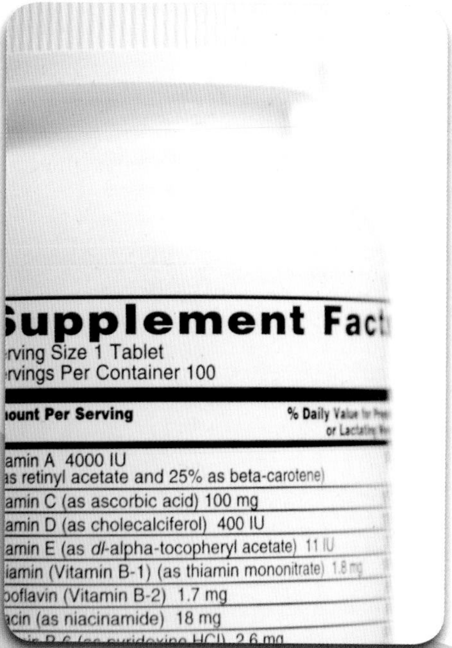

Look at the products around your home. What kinds of mixtures, solutions, and compounds do you find?

GLOSSARY

atoms: the basic particles that all matter is made of.

chemical change: a process a substance undergoes when it is involved in a chemical reaction.

chemical formula: a way to identify a substance using letters and numbers instead of words.

chemical reaction: chemical bonding between two or more substances that changes the properties of the substances.

chemist: a person who studies the interactions of substances and the new substances created by those interactions.

compound: a chemical combination of two or more elements.

concentration: the amount of one substance in relation to another substance.

condensation: the process of changing from a gas to a liquid.

dissolved: evenly distributed throughout a solution.

electrons: particles with a negative charge that move around the outer part of an atom.

elements: pure substances that are made of only one kind of atom.

evaporation: the process of changing from a liquid to a gas.

matter: the substance of everything on Earth. Matter has mass and takes up space.

mixture: a combination of different substances in which each substance keeps its own properties.

molecules: groupings of atoms chemically bonded together.

neutrons: neutral particles in an atom's nucleus.

nucleus: the core structure of an atom.

physical change: a change in the shape or form of a substance that does not change the properties of the substance.

properties: characteristics of a substance.

protons: positively charged particles in an atom's nucleus.

saturated: unable to dissolve any more added substances.

solution: a mixture in which the molecules of one substance are evenly distributed among the molecules of another substance.

substances: matter that is always the same composition for every sample. Some substances are made of just one kind of atom but others have two or more kinds of atoms. Gold, salt, and water are all substances.

FURTHER READING

Mixtures and Solutions, by Molly Aloian. Crabtree, 2009.

Materials, by Denise Walker. Smart Apple Media, 2008.

Visionlearning, Matter.
http://www.visionlearning.com/library/module_viewer.php?mid=49

Rader's chem4kids.com, Solutions and Mixtures.
http://www.chem4kids.com/files/matter_solution.html

ADDITIONAL NOTES

Page 11: Crystals formed.

Page 18: No. Magnets do not attract sugar.

Page 20: Iced tea and hot tea are made by mixing two or more substances. So are sugar, bronze, and brass. The substances in iced tea and hot tea still have their original properties. The substances in sugar, bronze, and brass do not have their original properties. The sugar, bronze, and brass have different properties. Iced tea and hot tea are liquids. Sugar, bronze, and brass are solids.

Page 23: Water melts at 32°F (0°C). It is colorless and tasteless.

Page 25: No, the sugar does not share electrons with the tea water. Electrons are shared only when there is a chemical change.

Page 26: Mixing sugar into water causes a physical change. The substance is still made of sugar and water. No new substance is formed.

Page 27: No. When the tea was made, no light was given off. There was no smell or sound. The tea still looks and acts like water, and it has the same properties as water.

Page 30: Children need less medicine because their bodies are smaller than adult bodies. Also, their body chemistry is different from adult body chemistry.

Page 31: The substances that make up the concrete are no longer in their original form.

Page 32: Sugarless iced tea is a solution of tea and water. Adding sugar makes it a solution of tea, water, and sugar.

Page 33: The sugar crystals have separated into individual molecules, which are too small to see.

Page 34: You could put the same amount of sugar in two cups and stir one cup but not the other. You could compare the amount of sugar you can dissolve in iced tea to the amount of sugar you can dissolve in hot tea.

Page 35: The liquid with two spoonfuls of sugar would have the higher concentration of sugar. The concentrations of the sugar in the cups of tea in the iScience Puzzle are the same.

Page 36: You could experiment with equal amounts of sugar crystals of different sizes to see which would dissolve slowest. You could also test how much sugar of different crystal sizes would dissolve into identical cups of tea.

Page 37: You would see sugar crystals at the bottom of the glass because no more sugar would dissolve.

Page 38: Magnets could not be used because they are not attracted to sugar. Sugar molecules are too small to be picked out with your fingers. Caption question: The filter will not trap sugar because the sugar is dissolved.

Page 39: The salt molecules formed crystals again and collected on the yarn. The water evaporated. You can separate the substances in some solutions by evaporating the water and collecting the dissolved substance left behind. The water would evaporate. The tea solid and sugar would be left behind.

Page 41: The water in the mixture could be evaporated as a gas and collected as a liquid. This process would leave the solid tea and sugar behind. The ingredients could be measured and then mixed together again.

INDEX